SpringerBriefs in Electrical and Computer Engineering

For further volumes:
http://www.springer.com/series/10059

Samantha Yoder · Mohammed Ismail
Waleed Khalil

VCO-Based Quantizers Using Frequency-to-Digital and Time-to-Digital Converters

 Springer

Samantha Yoder
The ElectroScience Laboratory
The Ohio State University
Columbus, OH 43212, USA
yoder.164@osu.edu

Waleed Khalil
The ElectroScience Laboratory
The Ohio State University
Columbus, OH 43212, USA
khalil@ece.osu.edu

Mohammed Ismail
Department of Electrical
and Computer Engineering
The Ohio State University
Columbus, OH 43210-1272, USA
ismail@ece.osu.edu

ISSN 2191-8112   e-ISSN 2191-8120
ISBN 978-1-4419-9721-0   e-ISBN 978-1-4419-9722-7
DOI 10.1007/978-1-4419-9722-7
Springer New York Dordrecht Heidelberg London

Library of Congress Control Number: 2011934750

Springer is part of Springer Science+Business Media (www.springer.com)

Preface

Traditional analog-to-digital converters (ADCs) face many design challenges as technology scales. A few of these challenges are (1) voltage dynamic range decreases making it difficult to accurately quantize in the voltage domain (2) architecture contains many analog components which are challenging to design in deep submicron complementary metal oxide semiconductor (CMOS) processes. Voltage-controlled oscillator (VCO)-based ADCs are gaining popularity due to the highly digital architecture and improved timing resolution in deep submicron CMOS processes.

This book presents a theoretical and modeling approach to understanding the VCO-based quantizer. Two digital time quantizer architectures are reviewed: one using a frequency-to-digital converter (FDC) and the other using a time-to-digital converter (TDC). The TDC architecture is new to the application of the VCO-based quantizer.

Chapter 1 provides an introduction including the motivation for this topic, background on the subject, and goals of this work.

Chapter 2 provides an introduction and theoretical analysis of the FDC and TDC VCO-based quantizer. Theoretical equations are developed to determine the resolution of the quantizers and verified through a VerilogA model.

Chapter 3 provides modeling and analysis of circuit nonidealities of the VCO-based quantizer. These nonidealities are added to the VerilogA model and theoretical equations derived to verify the effects on both the FDC and TDC architecture.

Chapter 4 provides some final thoughts and analysis on the FDC and TDC VCO-based quantizer.

Chapter 5 concludes the book.

Columbus, OH, USA

Samantha Yoder
Mohammed Ismail
Waleed Khalil

Contents

Abbreviations

ADC	Analog-to-digital converter
CMOS	Complementary metal oxide semiconductor
CP	Charge pump
DAC	Digital-to-analog converter
DEM	Dynamic element matching
DFF	D flip-flop
$\Delta\Sigma$	Delta-sigma
FDC	Frequency-to-digital converter
FFT	Fast Fourier transform
MB	Multibit
NTF	Noise transfer function
OSR	Oversampling ratio
PLL	Phase-locked loop
PSD	Power spectral density
PWM	Pulse width modulation
RVCO	Ring VCO
SNR	Signal-to-noise ratio
SNDR	Signal-to-noise plus distortion ratio
STF	Signal transfer function
TDC	Time-to-digital converter
THD	Total harmonic distortion
VCO	Voltage-controlled oscillator

Chapter 1
Introduction

In conventional analog-to-digital converters (ADCs), the input voltage signal is sampled and quantized in the voltage domain using an analog processing chain. Voltage-controlled oscillator (VCO)-based ADCs differ from traditional ADCs since the analog input voltage is first converted to timing information, via the VCO, and then quantized in time, see Fig. 1.1. The voltage to time conversion occurs since the VCO frequency is proportional to the input voltage, Fig. 1.2. The VCO output can then be quantized using a digital time quantizer.

The concept of VCO-based ADCs came about in 1999 and has recently gained popularity due to its highly digital implementation which takes advantage of deep submicron complementary metal oxide semiconductor (CMOS) processes. The VCO-ADC architecture reduces if not eliminates the need for complex high-performance analog circuits which are becoming more challenging as technology advances.

1.1 Motivation

Consider the typical ADC shown in Fig. 1.3. This ADC contains primarily analog circuit components which quantize the input signal in the voltage domain. As technology scales, voltage dynamic range decreases making it more difficult to quantize in the voltage domain. Alternatively, time resolution is increasing as technology scales, Fig. 1.4.

The VCO-based quantizer has been gaining popularity since it takes advantage of the timing resolution and also lends itself to a highly digital implementation and processing techniques.

1.2 Background

When the concept of the VCO-based quantizer was first introduced in 1999, it was utilized in a $\Delta\Sigma$ ADC to improve resolution, Fig. 1.5. The time quantizer used in this architecture was a simple digital reset counter, which would count the VCO

S. Yoder et al., *VCO-Based Quantizers Using Frequency-to-Digital and Time-to-Digital Converters*, SpringerBriefs in Electrical and Computer Engineering, DOI 10.1007/978-1-4419-9722-7_1, © Springer Science+Business Media, LLC 2011

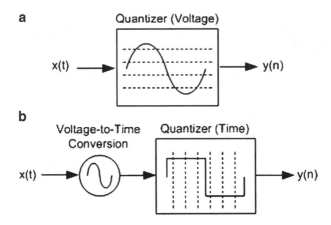

Fig. 1.1 (a) Typical ADC (voltage quantizer). (b) VCO-based ADC (time quantizer)

Fig. 1.2 Voltage to time conversion via the VCO

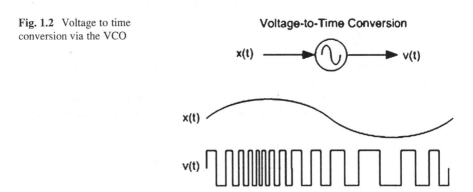

transitions within a sampling period to give a representation of the analog input voltage to the VCO, Fig. 1.6. It was derived that the resolution increased as a function of the VCO tuning frequency and oversampling ratio (OSR). Also the use of multiphases of the VCO could further enhance the resolution. As opposed to traditional voltage-based quantizers, it was shown that the VCO-based quantizer provided inherent first-order noise shaping, giving one more order of noise shaping to that of a traditional $\Delta\Sigma$ ADC.

The architecture showed promising results but suffered from harmonic distortion caused by nonlinearity of the VCO frequency tuning curve and errors due to missing "VCO transitions/counts" due to the reset operation of the counter [1]. A challenge in this design is the feedback digital-to-analog converter (DAC) that can be quite complex since requirements for nonlinearity and mismatch lead to complicated design techniques and dynamic element matching (DEM), which leads to primarily single-bit $\Delta\Sigma$ architectures. In the case of using the VCO-based quantizer

Fig. 1.3 Flash ADC

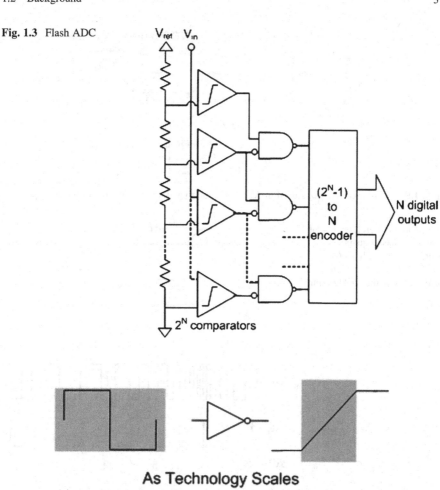

Fig. 1.4 As technology scales voltage resolution decreases while timing resolution increases

of Fig. 1.6, the output will be multibit (MB) and thus requires a complicated MB feedback DAC. In 2000, the design was re-architected to eliminate the feedback DAC by the use of a frequency detector in the feedback loop [2]. This design eliminated the DEM and mismatch errors; however, this approach did not make any further progress possibly due to the lack of interest in the subject. In 2006, the idea of using a multiphase VCO to improve the quantizer resolution was adopted.

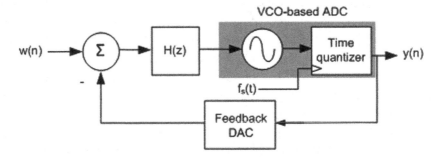

Fig. 1.5 VCO-based quantizer used in ΔΣ ADC loop to add an extra degree of noise shaping

Fig. 1.6 First architecture of
the VCO-based quantizer
implemented with a VCO and
counter to count frequency
transitions of the VCO output

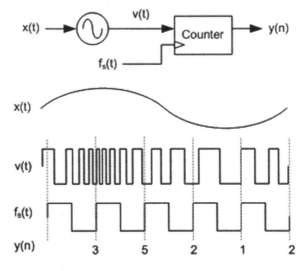

The design used a ring VCO to provide multiphase outputs. Each output requires a counter and the result is summed to improve the resolution, Fig. 1.7 [3].

Issues with the VCO nonlinearity still remain an issue and in 2007 the VCO was implemented as a fully differential architecture to cancel the even order harmonics due to the nonlinearity, Fig. 1.8 [4].

In 2007, a new architecture was introduced replacing the counter with a frequency-to-digital converter (FDC). The FDC gives a quantized representation of the VCO phase, Fig. 1.9, while still providing first-order noise shaping and eliminating the errors caused by the reset operation of the counter. Further advantages of this architecture include inherent DEM which was shown by its use in a ΔΣ modulator. It was also shown that the distortion due to VCO nonlinearity could be suppressed by the loop filter gain in the ΔΣ modulator; however, it still limits the signal-to-noise plus distortion ratio (SNDR) [5, 6].

Around 2008–2009, a new architecture was introduced which used time interleaving of the VCO-ADCs, Fig. 1.10, to develop a bandpass ADC [7, 8].

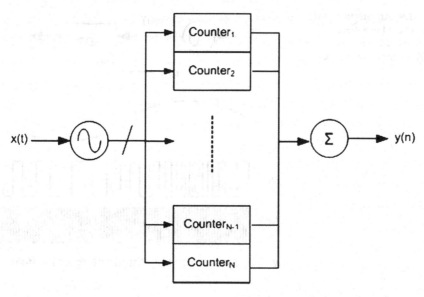

Fig. 1.7 VCO-based quantizer based on a multiphase VCO and multiple counters to improve resolution

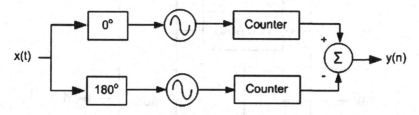

Fig. 1.8 VCO-based quantizer employing differential architecture to reduce VCO nonlinearity distortion

Around the same time another architecture was introduced to increase the noise shaping characteristics of the VCO-ADC without the complicated analog circuits and feedback DAC involved in the $\Delta\Sigma$ modulator. The architecture uses ideas of $\Delta\Sigma$ modulators but places the VCO at the front end, removed from the feedback path, Fig. 1.11. This allows a highly digital implementation, with exception of the charge pump (CP), and still provides a higher degree of noise shaping [9].

From 2009 and on, several ideas have been introduced to reduce VCO nonlinearity which remains the main bottleneck to these types of quantizers. Several architectures use the pseudodifferential architecture which cancels even order harmonics, and in addition use digital calibration techniques to cancel other harmonics [10–13]. Digital calibration which includes temperature compensation is presented in ref. [14]. Most recently, Hernández et al. [15] uses a pulse width modulation (PWM) precoding before the VCO to reduce nonlinearity distortion.

Fig. 1.9 Architecture of the
VCO-based quantizer
implemented with a VCO and
FDC to quantizer the VCO
phase

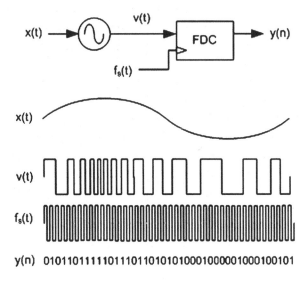

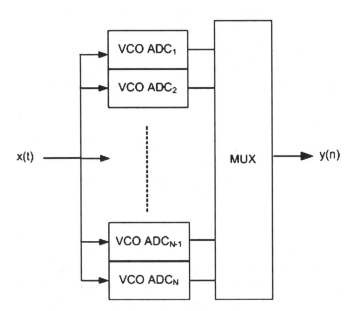

Fig. 1.10 Time interleaved VCO-based quantizer to increase the order of noise shaping

Recent architectures use some of the techniques described above such as: use of
a multiphase VCO and multiple counters/FDC's; linearization techniques such as
calibration, differential architectures, or use of a $\Delta\Sigma$ modulator. Successful
prototypes have been built in standard CMOS technologies to verify circuit performance [4–6, 8, 9, 11, 12].

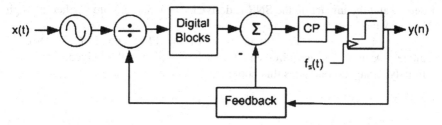

Fig. 1.11   VCO-based ADC with second-order noise shaping

1.3   Goals of this Work

This work introduces a new architecture for the VCO-based quantizer. The new architecture consists of a VCO and time-to-digital converter (TDC) to quantize the VCO signal. TDCs have been traditionally used in phase-locked loops (PLLs) to quantize the VCO phase error but have not been applied to VCO-based ADCs.

To provide a comparison, the TDC VCO-based quantizer is compared against the FDC VCO-based quantizer. Theoretical analysis and modeling of both quantizers is presented. Comparison of these digitization techniques will take into account VCO nonlinearity, phase noise, and sampling clock jitter. These nonidealities will be added to the model and theoretical equations will be derived to verify the effects on each quantizer. Although the FDC has been widely adopted due to its inherent first-order noise shaping characteristic, the noise shaping is shown to degrade in the presence of phase noise and clock jitter [16]. The new architecture may provide different advantages over the FDC in terms of design parameters and circuit performance. A final comparison on the performance of each quantizer and the circuit requirements for both will be presented.

1.4   Organization

Chapter 2 introduces the VCO-based quantizer. The VCO-based quantizer is described and analyzed in detail for both the FDC and TDC quantization method. Theoretical equations are derived to determine the quantizer resolution [signal-to-noise ratio (SNR)]. The SNR derivation for the FDC matches what has been found in the literature [16]. The TDC VCO-based quantizer SNR is derived for the first time. These theoretical equations are verified through VerilogA simulation and closely match the theoretical values. The VerilogA models are used in Chap. 3 to compare the performance of the FDC vs. TDC VCO-based quantizers in the presence of circuit nonidealities.

Chapter 3 describes the limitations of the VCO-based quantizer in the presence of circuit nonidealities such as VCO nonlinearity, VCO phase noise, and sampling clock jitter. Modeling of the nonidealities and theoretical analysis on the effects is presented and the impact on the SNR is derived. The equations are verified through a VerilogA model.

Chapter 4 includes a final comparison of the FDC and TDC VCO-based quantizer. Design issues/requirements for each quantizer will be given.

Finally, Chap. 5 concludes this book.

References

1. Iwata A, Sakimura N, Nagata M, Morie T (1999) The architecture of delta sigma analog-to-digital converters using a voltage-controlled oscillator as a mulibit quantizer. IEEE Trans Circ Syst II: Anal Dig Signal Process 46(7):941–945
2. Naiknaware R, Tang H, Fiez TS (2000) Time-referenced single-path multi-bit $\Delta\Sigma$ ADC using a VCO-based quantizer. IEEE Trans Circ Syst II: Anal Digit Signal Process 47(7):596–602
3. Kim J, Cho S (2006) A time-based analog-to-digital converter using a multi-phase voltage controlled oscillator. In: IEEE International Symposium on Circuits and Systems, 2006 (ISCAS 2006), Island of Kos, pp 3934–3937
4. Tritschler A (2007) A continuous time analog-to-digital converter with 90 μW and 1.8 μV/LSB based on differential ring oscillator structures. In: IEEE International Symposium on Circuits and Systems, 2007 (ISCAS 2007), 27–30 May 2007, pp 1229–1232
5. Straayer MZ, Perrott MH (2007) A 10-bit 20 MHz 38 mW 950 MHz CT $\Sigma\Delta$ ADC with a 5-bit noise-shaping VCO-based quantizer and DEM circuit in 0.13 nm CMOS. In: IEEE Symposium on VLSI Circuits, 14–16 June 2007, pp 246–247
6. Straayer MZ, Perrott MH (2008) A 12-bit 10-MHz bandwidth, continuous-time $\Delta\Sigma$ ADC with a 5-bit 950-MSs VCO-based quantizer. IEEE J Solid-State Circ 43(4):805–814
7. Yoon Y-G, Kim J, Jang T-K, Cho SH (2008) A time-based bandpass ADC using time-interleaved voltage-controlled oscillators. IEEE Trans Circ Syst I 55(11):3571–3581
8. Yoon Y-G, Cho SH (2009) A 1.5-GHz 63 dB SNR 20 mW direct RF sampling bandpass VCO-based ADC in 65 nm CMOS. In: Symposium on VLSI Circuits 2009, Kyoto, Japan, 16–18 June 2009, pp 270–271
9. Park M, Perrot M (2009) A VCO-based analog-to-digital converter with second-order sigma-delta noise shaping. In: IEEE International Symposium on Circuits and Systems (ISCAS), 2009, Taipei, 24–27 May 2009, pp 3130–3133
10. Daniels J, Dehaene W, Steyaert M (2010) All-digital differential VCO-based A/D conversion. In: Proceedings of 2010 IEEE International Symposium on Circuits and Systems (ISCAS), Paris, 30 May 2010–2 June 2010, pp 1085–1088
11. Daniels J, Dehaene W, Steyaert M (2010) A 0.02 mm^2 65 nm CMOS 30 MHz BW all-digital differential VCO-based ADC with 64 dB SNDR. In: 2010 IEEE Symposium on VLSI Circuits (VLSIC), Digest of Technical Papers, Honolulu, pp 155–156
12. Taylor G, Galton I (2010) A mostly-digital variable-rate continuous-time delta-sigma modulator ADC. IEEE J Solid State Circ 45(12):2634–2646
13. Venkatram H, Inti R, Un-Ku M (2010) Least mean square calibration method for VCO non-linearity. In: International Conference on Microelectronics (ICM), 2010, Cairo, 19–22 Dec 2010, pp 1–4
14. Negreiros M, Carro L, Cassel G (2010) All digital ADC with linearity correction and temperature compensation. IEEE Instrument and Measurement Technology Conference (I2MTC) 2010, Austin, TX, 3–6 May 2010, pp 147–152
15. Hernández L, Patón S, Prefasi E (2011) VCO-based sigma delta modulator with PWM precoding. Electronic Lett 47(10):588–589
16. Kim J et al (2010) Analysis and design of voltage-controlled oscillator based analog-to-digital converter. IEEE Trans Circ Syst I Reg Pap 57(1):18–30

Chapter 2
VCO-Based Quantizer

In this chapter a detailed analysis on the operation and architecture of the voltage-controlled oscillator (VCO)-based ADC is presented. The VCO-based quantizer is analyzed for two different architectures, one using a frequency-to-digital converter (FDC) the other a time-to-digital converter (TDC). Theoretical equations are derived to determine the resolution of these quantizers and verified through a VerilogA model.

2.1 VCO Operation

An ideal VCO output can be expressed as a sinusoidal signal with amplitude and phase as shown in (2.1) [1]:

$$\text{VCO}_{\text{out}}(t) = A_{\text{vco}} \sin \theta_{\text{vco}}(t), \tag{2.1}$$

$$\theta_{\text{vco}}(t) = 2\pi \int_0^\infty f_{\text{vco}}(\tau) d\tau = 2\pi \int_0^\infty f_c + K_v x(\tau) d\tau. \tag{2.2}$$

The phase of the VCO (2.2) is the integral of the VCO frequency which depends on VCO characteristics such as the free running center frequency, f_c, VCO gain, K_v, and input signal to the VCO, i.e., the control voltage $x(t)$. The block diagram of the VCO is shown in Fig. 2.1.

The voltage dynamic range of $x(t)$ is converted to a frequency dynamic range via the VCO, which depends on the control voltage, $x(t)$, along with the VCO gain, K_v. This is shown graphically in Fig. 2.2, where the VCO output frequency range is $f_c \pm \Delta f$, with $\Delta f = K_v x(t)$.

Next the methods of quantizing the VCO output signal are discussed. The FDC measures the phase difference between samples to arrive at the VCO frequency,

S. Yoder et al., *VCO-Based Quantizers Using Frequency-to-Digital and Time-to-Digital Converters*, SpringerBriefs in Electrical and Computer Engineering, DOI 10.1007/978-1-4419-9722-7_2, © Springer Science+Business Media, LLC 2011

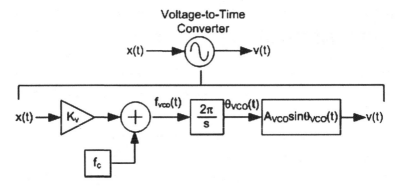

Fig. 2.1 VCO block diagram

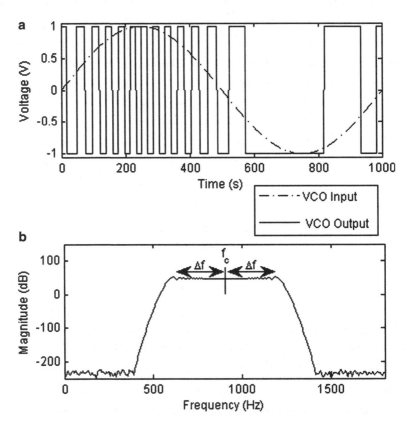

Fig. 2.2 (a) Buffered VCO output vs. sinusoidal VCO input signal (time domain). (b) VCO output with frequency range: $f_c \pm \Delta f$ (frequency domain)

Fig. 2.3 VCO-based
quantizer. (**a**) FDC: measures
phase difference. (**b**) TDC:
measures period and inverts

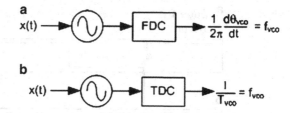

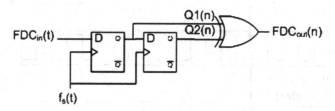

Fig. 2.4 FDC block diagram

while the TDC measures the period of the VCO which is inverted to get the VCO
frequency, Fig. 2.3.

2.2 FDC VCO-Based Quantizer

The architecture of the FDC consists of two D flip-flops (DFFs) and one XOR gate
as shown in Fig. 2.4 [2].

Consider the waveforms shown in Fig. 2.5. The FDC operation is as follows:
At the rising clock edge the first DFF output, Q1, samples the input signal. The value
of the second DFF output, Q2, takes on the value of Q1 before that rising clock
edge. The output of the FDC, the XOR gate, determines if there is a difference
between two consecutive samples. Consider the first occurrence when the FDC input
changes from 1 to 0. At the rising clock edge after this transition, the first DFF
output, Q1, changes from 1 to 0. The value of the second DFF output, Q2, takes on
the value of Q1 before that rising clock edge – 1. The XOR gate sees Q1 = 0 and
Q2 = 1 and it outputs a value of 1. Thus, the FDC operates as an edge detector, with
some delay or error when the transition is detected, this can be up to one clock
period.

The architecture of the FDC VCO-based quantizer is shown in Fig. 2.6 [3].
This architecture consists of a ring VCO to provide a multiphase output, and
several FDC's operating in parallel which are summed together to form a multibit

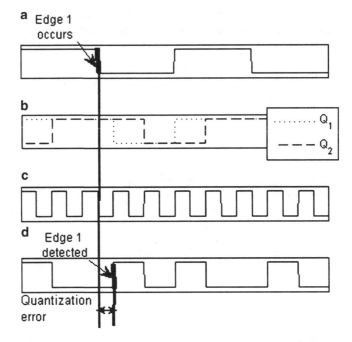

Fig. 2.5 FDC example waveforms. (**a**) FDC input; (**b**) output of sampled DFF 1 and 2; (**c**) sampling clock $f_s(t)$; (**d**) FDC output

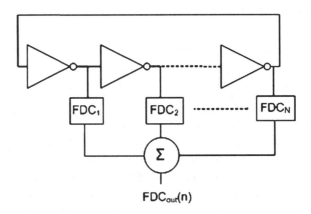

Fig. 2.6 Multibit FDC VCO-based quantizer

quantizer. The VCO phase taps are delayed from one another by $2\pi/N$, N being the number of stages. Thus, there are multiple reads of the signal within a single cycle to improve the resolution. The single-bit and multibit operations are shown in Fig. 2.7.

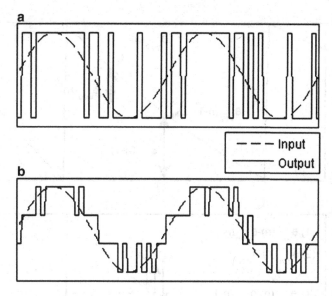

Fig. 2.7 FDC VCO-based quantizer input vs. output (time domain): (**a**) single bit (**b**) multibit

2.2.1   Linear Modeling and Analysis

The FDC VCO-based quantizer is modeled and analyzed using linear modeling to determine the resolution. As stated before, the FDC differentiates the VCO phase to get the VCO frequency. Figure 2.8 shows the VCO phase which is sampled at two consecutive samples denoted as $(n-1)$ and (n). θ_{vco} is the actual VCO phase and ϕ_{vco} is the phase error which is the difference between what the actual VCO phase is and when it last accumulated multiple of π. When the VCO phase accumulates π, the VCO output will undergo transition which is detected by the FDC at the next sampling instance.

The output of the FDC can be written as the phase difference between samples (2.3) using the relationships shown in Fig. 2.8. The equation includes the multibit operation of using "N" FDCs, similar analysis is given in ref. [4].

$$\text{FDC}_{\text{out}}(n) = \frac{N}{\pi}[\theta_{\text{vco}}(n) - \theta_{\text{vco}}(n-1) + \phi_{\text{vco}}(n-1) - \phi_{\text{vco}}(n)]. \qquad (2.3)$$

The VCO phase given in (2.2) is sampled and can be written as (2.4):

$$\theta_{\text{vco}}(n) = 2\pi \int_{(n-1)T_s}^{nT_s} f_c + K_v x(\tau) d\tau = 2\pi \sum_{i=1}^{n} T_s[f_c + K_v x(i)]. \qquad (2.4)$$

The FDC output equation is simplified in terms of the phase difference as shown in (2.5) and (2.6) to arrive at the output equation of the FDC (2.7).

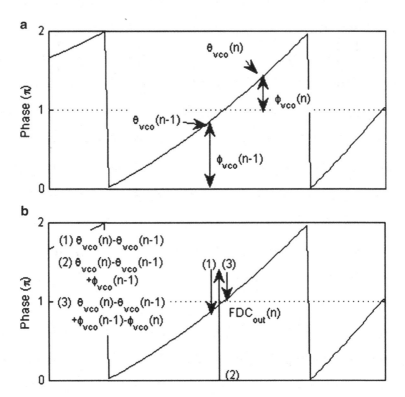

Fig. 2.8 Quantization of VCO phase by the use of FDC: (**a**) phase and phase error for two consecutive samples; (**b**) FDC output based on phase differences

$$\theta_{\text{vco}}(n) - \theta_{\text{vco}}(n-1) = 2\pi \sum_{i=1}^{n} T_s[f_c + K_v x(i)] - 2\pi \sum_{i=1}^{n-1} T_s[f_c + K_v x(i)], \quad (2.5)$$

$$\theta_{\text{vco}}(n) - \theta_{\text{vco}}(n-1) = 2\pi T_s[f_c + K_v x(n)], \quad (2.6)$$

$$\text{FDC}_{\text{out}}(n) = 2NT_s[f_c + K_v x(n)] + \frac{N}{\pi}[\phi_{\text{vco}}(n-1) - \phi_{\text{vco}}(n)]. \quad (2.7)$$

Using Z-transformations, the output of the FDC in terms of the frequency domain is written as (2.8):

$$\text{FDC}_{\text{out}}(z) = 2NT_s f_c + 2NT_s K_v X(z) + \frac{N}{\pi} \phi_{\text{vco}}(z) \left(\frac{1-z}{z} \right). \quad (2.8)$$

Based on (2.8), the signal transfer function (STF) and noise transfer function (NTF) for the FDC VCO-based quantizer is derived in (2.9) and (2.10).

$$\text{STF} = \frac{\text{FDC}_{\text{out}}(z)}{X(z)} = 2NT_sK_v, \qquad (2.9)$$

$$\text{NTF} = \frac{\text{FDC}_{\text{out}}(z)}{\phi_{\text{vco}}(z)} = \left(\frac{1-z}{z}\right)\frac{N}{\pi}. \qquad (2.10)$$

From the derivations, the input signal is scaled while the quantization noise is first-order shaped. To derive the SNR, the power of the signal and the power of the noise at the output of the FDC are calculated. Consider the input to the VCO as a sine wave (2.11), then the power of the output signal is given as follows (2.12):

$$x(t) = A_m \sin(2\pi f_m t), \qquad (2.11)$$

$$P_s = \left(\frac{2NT_sK_vA_m}{\sqrt{2}}\right)^2. \qquad (2.12)$$

The power of the quantization noise is derived next. The quantization noise is assumed to be white and thus the level of noise is constant across all frequencies and equal to (2.13) using the two-sided definition of power [5, pp. 532–533].

$$\phi_{\text{vco}}(n) = \frac{\Delta}{\sqrt{12f_s}} = \frac{\pi}{N\sqrt{12f_s}}, \qquad (2.13)$$

where Δ is the difference between quantization levels, in the FDC case this is equal to π and reduced by N when using a multibit architecture [6]. The noise is filtered through the NTF. For this analysis, the NTF is given in the frequency domain (2.14):

$$\text{NTF} = \left(\frac{1-z}{z}\right)\frac{N}{\pi}\Bigg|_{z=e^{j2\pi(f/f_s)}} = N\frac{e^{-j2\pi(f/f_s)}-1}{\pi}$$

$$= -N\frac{2j}{\pi}e^{-j\pi(f/f_s)}\frac{e^{j\pi(f/f_s)}-e^{-j\pi\pi(f/f_s)}}{2j}. \qquad (2.14)$$

Using Euler's formula, the magnitude of the NTF assuming that the sampling frequency is much greater than the bandwidth is given as follows (2.15):

$$|\text{NTF}| = N\frac{2}{\pi}\sin\left(\pi\frac{f}{f_s}\right) \approx N\frac{2}{\pi}\frac{\pi f}{f_s} = N\frac{2f}{f_s}. \qquad (2.15)$$

The in-band power given a bandwidth equal to f_b is given in (2.16) using (2.13) and (2.15):

$$P_N = 2\int_0^{f_b} \left(\frac{2f}{f_s}\frac{\pi}{\sqrt{12f_s}}\right)^2 df = 8\frac{\pi^2 f_b^3}{36f_s^3} = \frac{\pi^2}{36OSR}. \qquad (2.16)$$

The power of the noise matches the general equation used for first-order noise shaping [5, p. 552]. The SNR is the power of the signal divided by the power of the noise, (2.12) divided by (2.16), given as follows (2.17):

$$SNR = 10\log_{10}\left(\frac{\left(2NT_sK_vA_m/\sqrt{2}\right)^2}{8(\pi^2 f_b^3/36f_s^3)}\right) = 10\log_{10}\left(\frac{9f_s^3(2NT_sK_vA_m)^2}{4\pi^2 f_b^3}\right). \qquad (2.17)$$

The maximum SNR of the FDC VCO-based quantizer depends on the number of FDCs, the ratio of the sampling frequency to the system bandwidth, and the VCO gain factor. This SNR value is consisted with what has previously been derived [4].

2.2.2   Model Verification Using VerilogA

To verify the above analysis and make further assessment, the FDC VCO-based quantizer is modeled in VerilogA and shown below.

```
//Begin VerilogA code: FDC VCO-based quantizer
`include "constants.vams"
`include "disciplines.vams"
`define PI 3.14159265

module fdc(in,clk,out);
input in, clk;
output out;
voltage in,clk,out;

parameter real        vdd=0.6,        // Positive Supply
                      vss=-0.6,       // Negative Supply
                      fc=500e6,       // VCO center frequency
                      Kv=250e6;       // VCO gain Hz/V
parameter             N=5;            // number of phase taps

// Define variables
real vout,fvco,phase, Ac, vmid;
real vcout[0:N-1];
integer q1[0:N-1],q2[0:N-1];
integer i,count;
```

```
analog begin
// Initialized Parameters
        @ (initial_step) begin
                Ac=(vdd-vss)/2.0;      // VCO Amplitude
                vmid=(vdd+vss)/2.0;    // Midrail Voltage
        end
// Ideal VCO frequency and phase equations
        fvco=Kv*V(in)+fc;      // VCO Frequency
        phase=2.0*`PI*idtmod(fvco,0,1,-0.5);       // VCO Phase
        for(i=0;i<N;i=i+1) begin
// Multiphase VCO output "N" phase taps
                vcout[i]=vmid+Ac*sin(phase+i*2.0*`PI/N);
// Buffer multiphase VCO outputs
                if (vcout[i]>=vmid) vcout[i]=vdd;
                if (vcout[i]<vmid) vcout[i]=vss;
                end
// "N" FDCs samples at the rising clock edge
        @ (cross(V(clk)-vmid,1)) begin
                count=0;
                for(i=0;i<N;i=i+1) begin
                q2[i]=q1[i];  // DFF2=DFF1
                q1[i]=vcout[i]>=vmid;// DFF1=buffered VCO output
// XOR Q1 and Q2 - add for multibit output
                if (q1[i] != q2[i]) count=count+1;
                end
        end
        vout= count    // Output
        V(out) <+ vout;
end
endmodule
//End VerilogA code: FDC VCO-based quantizer
```

The input to the VerilogA model is the clock signal and the input control voltage to the VCO. The VCO is modeled using the ideal relationships and buffered. At the rising clock edge, each FDC operates on its own buffered VCO phase tap. The output of the model is the sum of the FDCs.

To verify the theoretical analysis, multiple simulations of the VerilogA code are run varying different circuit parameters, Fig. 2.9. The theoretical analysis shows good correlation with the VerilogA model.

2.3 TDC VCO-Based Quantizer

The architecture of the TDC consists of a delay chain, several DFFs, and a digital pulse detector as shown in Fig. 2.10 [7].

Consider the waveforms shown in Fig. 2.11. The TDC operation is as follows: the input to the TDC is a pulse which will propagate through the delay chain and be sampled through the DFF's providing a snapshot of that pulse. Then using digital logic, the width of that pulse may be detected with some error which depends on the delay of the buffer (τ_{buff}) in the delay chain. The output of the pulse detector gives a

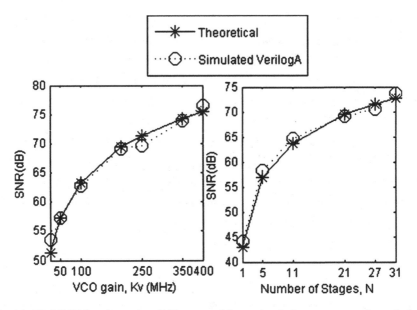

Fig. 2.9 FDC VCO-based quantizer SNR curves while varying design parameters: theoretical vs. simulated VerilogA model

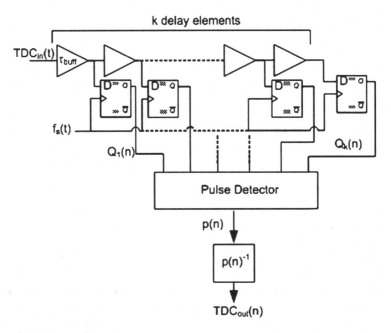

Fig. 2.10 TDC block diagram

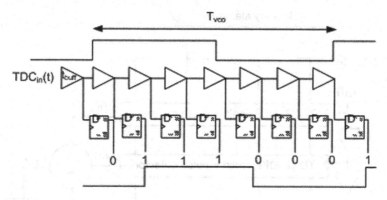

Fig. 2.11 Quantization of VCO period by the use of a TDC

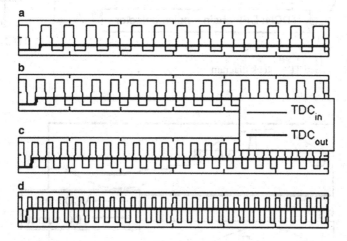

Fig. 2.12 TDC example waveform

representation of the period of that pulse and can be inverted to get the frequency; some results are shown in Fig. 2.12.

Like the FDC, the TDC can have "N" stages cascaded together, Fig. 2.13. The same delay chain can be used for each TDC; however, different sampling DFFs and pulse width detectors are needed. The clock signal is delayed to each TDC to cover a complete sampling cycle and the output of each TDC will be added to improve the resolution. This is the equivalent of increasing the sampling clock by "N". This is a slightly different approach to improve the dynamic range of the TDC than taken in ref. [7].

The architecture of the TDC VCO-based quantizer consists of any type of VCO and the TDC shown in Fig. 2.10. The resolution of the quantizer is improved by using multiple TDCs but only requires one-phase output of the VCO. The single TDC and multi-TDC VCO-based quantizer operations are shown in Fig. 2.14.

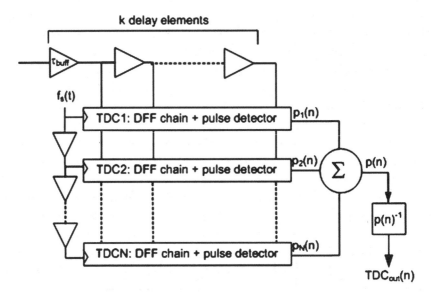

Fig. 2.13 Cascaded TDC block diagram

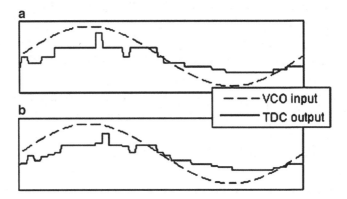

Fig. 2.14 TDC VCO-based quantizer input vs. output (time domain): (**a**) single TDC (**b**) multi-TDC

2.3.1   Linear Modeling and Analysis

The TDC VCO-based quantizer is modeled and analyzed using linear modeling to determine the resolution. As stated before, the TDC measures the period of the VCO and is inverted to get the VCO frequency. The detected pulse using the relationship shown in Fig. 2.11 is given in (2.18), including multiple TDCs.

$$p(n) = \frac{NT_{vco}(n)/2 + \sqrt{N}q(n)}{\tau_{buff}}. \tag{2.18}$$

The period of the VCO is divided by two since only the positive going pulse is detected, and it is quantized in terms of the buffer delay, τ_{buff}, with quantization noise, $q(n)$. The noise is assumed to be uncorrelated and adds as the square root of N, while the signal adds constructively. Since the frequency of the VCO is of interest, the pulse width is inverted to give the TDC output (2.19).

$$TDC_{out}(n) = \frac{1}{p(n)} = \frac{1}{(N/f_{vco}(n)2\tau_{buff}) + (\sqrt{N}q(n)/\tau_{buff})}. \tag{2.19}$$

The TDC output is simplified for analysis

$$TDC_{out}(n) = \frac{1}{(N/f_{vco}(n)2\tau_{buff}) + (\sqrt{N}q(n)/\tau_{buff})}$$
$$\cdot \frac{(N/f_{vco}(n)2\tau_{buff}) - (\sqrt{N}q(n)/\tau_{buff})}{(N/f_{vco}(n)2\tau_{buff}) - (\sqrt{N}q(n)/\tau_{buff})}, \tag{2.20}$$

$$TDC_{out}(n) = \frac{(N/f_{vco}(n)2\tau_{buff}) - (\sqrt{N}q(n)/\tau_{buff})}{(N/f_{vco}(n)2\tau_{buff})^2 + (\sqrt{N}q(n)/\tau_{buff})^2}. \tag{2.21}$$

Assuming the second term in the denominator of (2.21) is much smaller than the first, the TDC output can further be simplified (2.22):

$$TDC_{out}(n) \approx \frac{2\tau_{buff}f_{vco}(n)}{N} - \frac{4\sqrt{N}\tau_{buff}f_{vco}{}^2(n)q(n)}{N^2}. \tag{2.22}$$

Plugging in (2.2), the TDC output is approximately (2.23):

$$TDC_{out}(n) \approx \frac{2\tau_{buff}[f_c + K_v x(n)]}{N} - \frac{4\sqrt{N}\tau_{buff}[f_c + K_v x(n)]^2 q(n)}{N^2}. \tag{2.23}$$

One thing to note is that the quantization noise changes when the input signal changes. There is some correlation between the noise and input signal which is nonlinear. For simplification, this correlation is ignored and the final TDC output is given as follows (2.24):

$$\text{TDC}_{\text{out}}(n) \approx \frac{2\tau_{\text{buff}}[f_{\text{c}} + K_{\text{v}}x(n)]}{N} - \frac{4\sqrt{N}\tau_{\text{buff}}f_{\text{c}}^2 q(n)}{N^2}. \tag{2.24}$$

Based on (2.24), the STF and NTF for the TDC VCO-based quantizer are derived in (2.25) and (2.26).

$$\text{STF} = \frac{\text{TDC}_{\text{out}}(n)}{x(n)} = \frac{2\tau_{\text{buff}}K_{\text{v}}}{N}. \tag{2.25}$$

$$\text{NTF} = \frac{\text{TDC}_{\text{out}}(n)}{q(n)} = \frac{4\sqrt{N}\tau_{\text{buff}}f_{\text{c}}^2}{N^2}. \tag{2.26}$$

From the derivations, the input signal and noise are both scaled and dependant on some of the same factors. To derive the SNR, the power of the signal and the power of the noise at the output of the TDC are calculated. Consider the input to the VCO as a sine wave (2.11), then the power of the output signal is given in (2.27).

$$P_{\text{s}} = \left(\frac{2\tau_{\text{buff}}K_{\text{v}}A_{\text{m}}}{N\sqrt{2}}\right)^2. \tag{2.27}$$

The power of the quantization noise is derived next. The quantization noise is assumed to be white and thus the level of noise is constant across all frequencies and equal to (2.28) using the two-sided definition of power [5, pp. 532–533].

$$q(n) = \frac{\Delta}{\sqrt{12f_{\text{s}}}} = \frac{\tau_{\text{buff}}}{\sqrt{12Nf_{\text{s}}}}, \tag{2.28}$$

where Δ is the difference between quantization levels, in the TDC case this is equal to τ_{buff}. The effective sampling frequency is increased by N when using N TDCs to reduce the quantization noise. The noise is filtered through the NTF and the in-band power given a bandwidth equal to f_{b} is given in (2.29) and (2.30) using (2.26) and (2.28).

$$P_{\text{N}} = 2\int_0^{f_{\text{b}}} \left(\frac{4\sqrt{N}\tau_{\text{buff}}f_{\text{c}}^2}{N^2}\frac{\tau_{\text{buff}}}{\sqrt{12Nf_{\text{s}}}}\right)^2 df = 2\int_0^{f_{\text{b}}} \left(\frac{4\tau_{\text{buff}}^2 f_{\text{c}}^2}{N^2\sqrt{12f_{\text{s}}}}\right)^2 df, \tag{2.29}$$

$$P_{\text{N}} = \frac{32f_{\text{b}}\tau_{\text{buff}}^4 f_{\text{c}}^4}{N^4 12f_{\text{s}}}. \tag{2.30}$$

The SNR (2.31) is the power of the signal divided by the power of the noise, (2.27) divided by (2.30).

$$\mathrm{SNR} = 10\log_{10}\left(\frac{\left((2\tau_{\mathrm{buff}}K_vA_m)/(N\sqrt{2})\right)^2}{(32f_b\tau_{\mathrm{buff}}{}^4f_c^4)/(N^412f)}\right) = 10\log_{10}\left(\frac{3f_s(NK_vA_m)^2}{4\tau_{\mathrm{buff}}{}^2f_c^4f_b}\right). \quad (2.31)$$

The maximum SNR of the TDC VCO-based quantizer depends on the number of TDCs, the ratio of the OSR, the VCO gain factor, and also depends on the VCO center frequency and buffer delay. To the author's knowledge, this is the first time these equations have been derived. TDCs are traditionally used to detect the time or phase difference of two signals, but have not been used to detect the frequency of a signal.

2.3.2   Model Verification Using VerilogA

To verify the above analysis and make further assessment, the TDC VCO-based quantizer is modeled in VerilogA and shown below.

```
//Begin VerilogA code: TDC VCO-based quantizer
`include "constants.vams"
`include "disciplines.vams"
`define PI 3.14159265

module tdc(in, clk1,clk2,clk3,clk4,clk5 ,out);
input in,clk1,clk2,clk3,clk4,clk5,;
output out;
electrical in,clk1,clk2,clk3,clk4,clk5,out;

parameter real       tbuff= 33p,   // Buffer Delay
                     vdd = 0.6,    // Positive power supply
                     vss = -0.6,   // Negative power supply
                     fc=500M,      // VCO center frequency
                     Kv=250M;      // VCO gain factor Hz/V
parameter            N=5,          // Number of TDCs
                     n = 150;      // Length of delay chain

// Define variables
integer q1[0:n-1],q2[0:n-1],q3[0:n-1],q4[0:n-1],q5[0:n-1]
integer i,vs11,vs12,vs21,vs22,vs31,vs32,vs41,vs42,vs51,vs52;
real Ac,vmid,phase,fvco,vcout,vp1,vp2,vp3,vp4,vp5 ,vp, d[0:n-1];
```

```
analog begin
//Initialize parameters
      @(initial_step) begin
            Ac=(vdd-vss)/2.0;      // VCO amplitude
            vmid=(vdd+vss)/2.0;    // Midrail Voltage
      end
// Ideal VCO equations
      fvco=Kv*V(in)+fc;                  // VCO frequency
      phase=2.0*`PI*idtmod(fvco,0,1,-0.5);      // VCO phase
      vcout=vmid+Ac*sin(phase);    // VCO output
// Delay the VCO pulse along the delay chain
      @(timer(0,tbuff)) begin
            for(i=1;i<n;i=i+1) begin
                  d[n-i]=d[n-i-1];
            end
      d[0]=vcout;
      end
// Calculate the detected pulse width using TDC
// TDC1
      @(cross(V(clk1)-vmid,1)) begin
            for(i=0;i<n;i=i+1) begin      // Buffer VCO output
                  q1[i]=(d[i]>=vmid);
            end
            for(i=1;i<n;i=i+1) begin // Determine pulse width
                  if (q1[i]==1 && q1[i-1]==0) vs11=i;
                  if (q1[i]==0 && q1[i-1]==1) vs12=i;
            end
            vp1=abs(vs11-vs12);  // Pulse width
      end
// TDC2
      @(cross(V(clk2)-vmid,1)) begin
            for(i=0;i<n;i=i+1) begin
                  q2[i]=(d[i]>=vmid);
            end
            for(i=1;i<n;i=i+1) begin
                  if (q2[i]==1 && q2[i-1]==0) vs21=i;
                  if (q2[i]==0 && q2[i-1]==1) vs22=i;
            end
            vp2=abs(vs21-vs22);
      end
// TDC3
      @(cross(V(clk3)-vmid,1)) begin
            for(i=0;i<n;i=i+1) begin
                  q3[i]=(d[i]>=vmid);
            end
            for(i=1;i<n;i=i+1) begin
                  if (q3[i]==1 && q3[i-1]==0) vs31=i;
                  if (q3[i]==0 && q3[i-1]==1) vs32=i;
            end
            vp3=abs(vs31-vs32);
      end
```

```
// TDC4
        @(cross(V(clk4)-vmid,1)) begin
                for(i=0;i<n;i=i+1) begin
                        q4[i]=(d[i]>=vmid);
                end
                for(i=1;i<n;i=i+1) begin
                        if (q4[i]==1 && q4[i-1]==0) vs41=i;
                        if (q4[i]==0 && q4[i-1]==1) vs42=i;
                end
                vp4=abs(vs41-vs42);
        end
// TDC5
        @(cross(V(clk5)-vmid,1)) begin
                for(i=0;i<n;i=i+1) begin
                        q5[i]=(d[i]>=vmid);
                end
                for(i=1;i<n;i=i+1) begin
                        if (q5[i]==1 && q5[i-1]==0) vs51=i;
                        if (q5[i]==0 && q5[i-1]==1) vs52=i;
                end
                vp5=abs(vs51-vs52);
        end
        @(cross(V(clk1)-vmid,1)) begin
// Add TDC outputs
                vp=vp1+vp2+vp3+vp4+vp5;
        end
// Convert to frequency
V(out)<+1.0/vp;
end
endmodule
//End VerilogA code: TDC VCO-based quantizer
```

The input to the VerilogA model is the clock signals and the input to the VCO. The VCO is modeled using the ideal relationships, delayed along the chain, buffered, and then fed to each TDC (five TDCs in this case). At the rising clock edge of each TDC, the pulse width is determined. The output of the model is the inverted sum of the TDCs.

To verify the theoretical analysis, multiple simulations of the VerilogA code are run varying different circuit parameters, Fig. 2.15. The theoretical analysis shows good correlation with the VerilogA model with some slight variations due to approximation.

2.4 FDC vs. TDC Architecture

So far, theoretical modeling and analysis has been developed for both the FDC and TDC VCO-based quantizers. The theoretical analysis is verified through a VerilogA model. The VerilogA model will be used to make further comparisons between the two quantization methods. The VCO-based quantizers are designed to achieve an SNR = 60 dB. The design parameters are chosen to produce the desired SNR and outlined in Table 2.1, where the same baseline VCO is used in

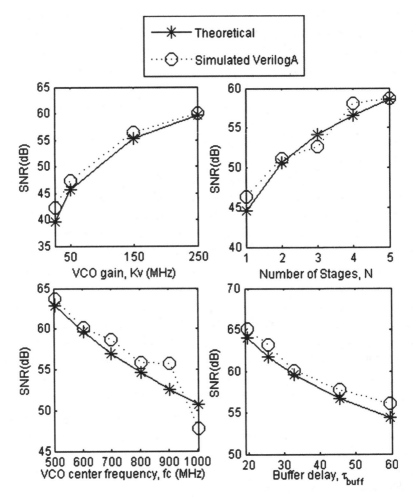

Fig. 2.15 TDC VCO-based quantizer SNR curves while varying design parameters: theoretical vs. simulated VerilogA model

Table 2.1 FDC/TDC design parameters to achieve SNR

Design parameters	FDC	TDC
f_s (GHz)	2	2
f_b (MHz)	10	10
A_m (V)	0.6	0.6
f_c (MHz)	500	500
K_v (MHz/V)	250	250
N	5	5
τ_{buff} (ps)	–	32.579

both quantizers. The theoretical SNR for this combination is shown in (2.32) and (2.33) for the FDC and TDC, respectively.

$$\text{SNR} = 10\log_{10}\left(\frac{9f_s^3(2NT_sK_vA_m)^2}{4\pi^2f_b^3}\right) = 60.11 \text{ dB}, \qquad (2.32)$$

Table 2.2 FDC/TDC
time domain simulation
parameters

Simulation parameters	
$V_{DD} - V_{SS}$ (V)	1.2
f_m (kHz)	500
T_{start} (µs)	2
T_{stop} (µs)	100
$T_{simulator}$ (ps)	250
Points	392,000

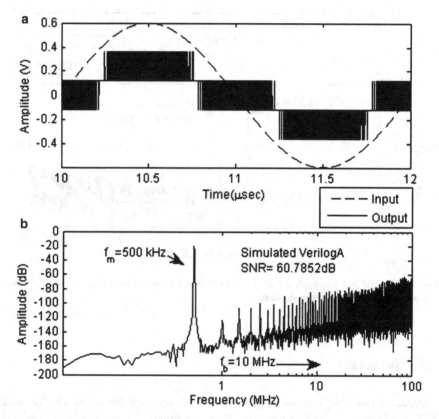

Fig. 2.16 Simulated VerilogA model FDC VCO-based quantizer: (**a**) input vs. output (time domain); (**b**) output (frequency domain)

$$\mathrm{SNR} = 10\log_{10}\left(\frac{3f_s(NK_vA_m)^2}{4\tau_{\mathrm{buff}}^2 f_c^4 f_b}\right) = 61.03 \text{ dB}. \qquad (2.33)$$

For simulation purpose, Table 2.2 is given.

The results for both VCO-based quantizers are shown in both time domain and frequency domain, where the output FFT is Hann windowed. The FDC results are shown in Fig. 2.16 and the TDC results are shown in Fig. 2.17. The simulated VerilogA SNR closely matches the theoretical SNR.

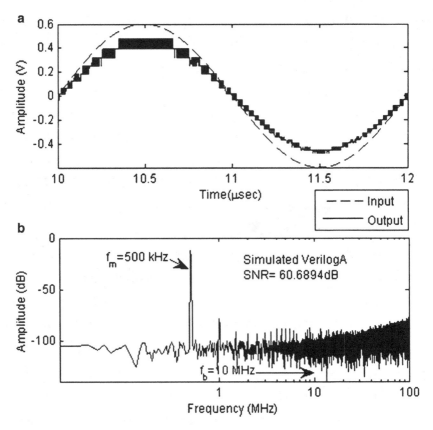

Fig. 2.17 Simulated VerilogA model TDC VCO-based quantizer: (**a**) input vs. output (time domain); (**b**) output (frequency domain)

2.5   Summary

Theoretical modeling and analysis of the VCO-based quantizer using an FDC and TDC has been developed to determine the resolution of the quantizers and verified through a VerilogA model. To make further comparison between the two quantization methods, the VerilogA models are used to achieve an SNR = 60 dB. The design parameters and characteristics for the two different quantizers are summarized in Table 2.3. Both systems are designed with the same sampling frequency and bandwidth. The VCO for both designs is the same and each quantizer consists of the same number of stages. The FDC has first-order noise shaping while the TDC does not. However, the TDC has two more SNR tuning knobs than the FDC.

Table 2.3 FDC/TDC VCO-based quantizer performance summary

Parameters	FDC VCO-based quantizer	TDC VCO-based quantizer
f_s, sampling frequency (GHz)	2	2
f_b, system bandwidth (MHz)	10	10
f_m $[0:f_b]$, VCO input signal frequency (kHz)	500	500
f_c, VCO center frequency (MHz)	500	500
K_v, VCO gain (MHz/V)	250	250
N, number of stages	5	5
Noise shaping	First order	None
SNR tuning knobs	$\uparrow(f_s, N, K_v)$	$\uparrow(f_s, N, K_v)\downarrow(\tau_{buff}, f_c)$
SNR (dB)	60.78	60.68

The FDC and TDC VCO-based quantizers will suffer from VCO nonidealities such as VCO nonlinearity and phase noise along with sampling clock jitter. The impact of these nonidealities on the VCO-based quantizer is analyzed and modeled in the next chapter.

References

1. Cao T, Wisland DT, Lande TS, Moradi F (2008) Low-voltage, low-power, and wide-tuning-range ring-VCO for frequency delta-sigma modulator. In: NORCHIP, 2008, Tallinn, 16–17 Nov 2008, pp 79–84
2. Wisland DT, Høvin ME, Lande TS, Saether T (1997) A narrow-band delta-sigma frequency-to-digital converter. In: 1997 IEEE International Symposium on Circuits and Systems, Hong Kong, 9–12 June 1997, pp 77–80
3. Straayer MZ, Perrott MH (2008) A 12-Bit 10-MHz bandwidth, continuous-time $\Delta\Sigma$ ADC with a 5-bit 950-MSs VCO-based quantizer. IEEE J Solid-State Circ 43(4):805–814
4. Kim J et al (2010) Analysis and design of voltage-controlled oscillator based analog-to-digital converter. IEEE Trans Circ Syst I: Reg Pap 57(1):18–30
5. Johns D, Martin K (1997) Analog integrated circuit design. Wiley, Canada
6. Wisland DT, Høvin ME, Lande TS (2002) A novel multi-bit parallel $\Delta\Sigma$ FM-to-digital converter with 24-bit resolution. Solid-State Circuits Conference, 2002. ESSCIRC 2002. Proceedings of the 28th European, 24–26 Sept. 2002, pp 687–690.
7. Tangudu J et al (2009) Quantization noise improvement of Time to Digital converter (TDC) for ADPLL. In: IEEE International Symposium on Circuits and Systems, 2009 (ISCAS 2009), Taipei, 24–27 May 2009, pp 1020–1023.

Chapter 3
Limitations of the VCO-Based Quantizer

In the following sections, both quantizers are compared in the presence of circuit nonidealities such as VCO nonlinearity, phase noise, and sampling clock jitter. These nonidealities are added to the VerilogA model, and theoretical equations are derived to verify the effects on each quantizer. Although the FDC has been widely adopted due to its inherent first-order noise shaping characteristic, the noise shaping is shown to degrade in the presence of phase noise and clock jitter. Other circuit nonidealities exist but are ignored in this analysis since these quantizers are highly digital circuits.

3.1 VCO Nonlinearity

The main bottleneck to these types of quantizers is the VCO nonlinearity. Nonlinearity in the VCO tuning curve will result in harmonic spurs in the output spectrum, Fig. 3.1. These spurs will degrade the signal-to-noise plus distortion ratio (SNDR) depending on the amount of nonlinearity in the tuning curve.

3.1.1 Modeling and Analysis

To analyze the effects of VCO nonlinearity, a ring VCO is designed for the VCO-based quantizer with a center frequency of 500 MHz and VCO tuning curve to satisfy the required 250 MHz/V gain. The simulated VCO tuning curve is shown in Fig. 3.2.

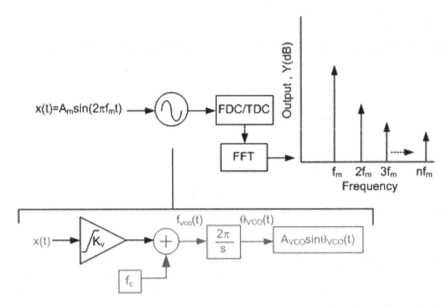

Fig. 3.1 VCO nonlinearity shown as harmonic spurs in the output frequency spectrum of the VCO-based quantizer

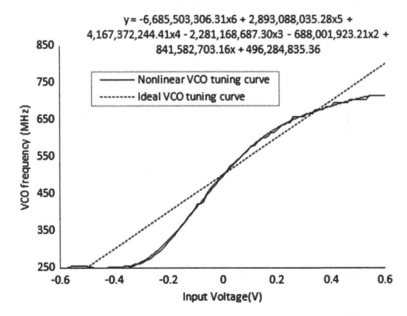

Fig. 3.2 VCO tuning curve: ideal vs. nonlinear tuning curve modeled with a sixth-order polynomial

The tuning curve is modeled using a sixth-order polynomial generalized in (3.1).

$$f_{vco}(x) = a_6 x(t)^6 + a_5 x(t)^5 + a_4 x(t)^4 + a_3 x(t)^3 + a_2 x(t)^2 + a_1 x(t) + a_0. \quad (3.1)$$

From Fig. 3.2, the variable a_n can be found. The VerilogA code for the nonlinear VCO is shown below. The VCO frequency is written as the nonlinear equation, with coefficients depending on the VCO tuning curve.

```
// Begin VerilogA code adjustment: Nonlinear VCO
      // Adjust VCO tuning curve with nonlinear parameters
      fvco=a6*pow(V(in),6)+a5*pow(V(in),5)+a4*pow(V(in),4)+a3*po
w(V(in),3)+a2*pow(V(in),2)  +a1*V(in)+a;
      phase=2.0*`PI*idtmod(fvco,0,1,-0.5);// VCO Phase
// End VerilogA code adjustment: Nonlinear VCO
```

To analyze the impact of the nonlinear tuning curve, consider the input to the VCO as a sine wave (3.2).

$$x(t) = A_m \sin 2\pi f_m(t) = A_m \sin \theta_m(t). \quad (3.2)$$

The VCO frequency in (3.1) can be rewritten using (3.2).

$$f_{vco}(x) = a_6 [A_m \sin \theta_m(t)]^6 + a_5 [A_m \sin \theta_m(t)]^5 + a_4 [A_m \sin \theta_m(t)]^4$$
$$+ a_3 [A_m \sin \theta_m(t)]^3 + a_2 [A_m \sin \theta_m(t)]^2 + a_1 A_m \sin \theta_m(t) + a_0. \quad (3.3)$$

Using trig identities, the VCO frequency equation can be expended (3.4)–(3.8).

$$a_2 [A_m \sin \theta_m(t)]^2 = a_2 A_m^2 \frac{1 - \cos 2\theta_m(t)}{2}, \quad (3.4)$$

$$a_3 [A_m \sin \theta_m(t)]^3 = a_3 A_m^3 \frac{3 \sin \theta_m(t) - \sin 3\theta_m(t)}{4}, \quad (3.5)$$

$$a_4 [A_m \sin \theta_m(t)]^4 = a_4 A_m^4 \frac{3 - 4 \cos 2\theta_m(t) + \cos 4\theta_m(t)}{8}, \quad (3.6)$$

$$a_5 [A_m \sin \theta_m(t)]^5 = a_5 A_m^5 \frac{10 \sin \theta_m(t) - 5 \sin 3\theta_m(t) + \sin 5\theta_m(t)}{16}, \quad (3.7)$$

$$a_6 [A_m \sin \theta_m(t)]^6 = a_6 A_m^6 \frac{10 - 6 \cos 2\theta_m(t) + 3 \cos 4\theta_m(t) - \cos 6\theta_m(t)}{32}. \quad (3.8)$$

The power at each frequency is combined to get the power of the signal (3.9) and power of the harmonic distortion terms (3.10)–(3.14).

$$P(f_{\mathrm{m}}) = a_1 A_{\mathrm{m}} + \frac{3}{4} a_3 A_{\mathrm{m}}^3 \frac{10}{16} a_5 A_{\mathrm{m}}^5, \tag{3.9}$$

$$P(2f_{\mathrm{m}}) = \frac{1}{2} a_2 A_{\mathrm{m}}^2 + \frac{4}{8} a_4 A_{\mathrm{m}}^4 \frac{6}{32} a_6 A_{\mathrm{m}}^6, \tag{3.10}$$

$$P(3f_{\mathrm{m}}) = \frac{1}{4} a_3 A_{\mathrm{m}}^3 + \frac{5}{16} a_5 A_{\mathrm{m}}^5, \tag{3.11}$$

$$P(4f_{\mathrm{m}}) = \frac{1}{8} a_4 A_{\mathrm{m}}^4 + \frac{3}{32} a_6 A_{\mathrm{m}}^6, \tag{3.12}$$

$$P(5f_{\mathrm{m}}) = \frac{1}{16} a_5 A_{\mathrm{m}}^5, \tag{3.13}$$

$$P(6f_{\mathrm{m}}) = \frac{1}{32} a_6 A_{\mathrm{m}}^6. \tag{3.14}$$

From the above analysis, the total harmonic distortion (THD) caused by the VCO nonlinear tuning curve is given as follows (3.15):

$$\mathrm{THD} = \frac{P(2f_{\mathrm{m}}) + P(3f_{\mathrm{m}}) + P(4f_{\mathrm{m}}) + P(5f_{\mathrm{m}}) + P(6f_{\mathrm{m}})}{P(f_{\mathrm{m}})}. \tag{3.15}$$

The SNDR is defined as the power of the signal divided by the power of the noise plus distortion (3.16).

$$\mathrm{SNDR} = 10\log_{10}\left(\frac{P_{\mathrm{s}}}{P_{\mathrm{N}} + P_{\mathrm{D}}}\right). \tag{3.16}$$

The equation is simplified in terms of the SNR, signal power, and THD, which will be a close approximation to the simulated SNDR.

$$\mathrm{SNDR} = 10\log_{10}\left(\frac{P_{\mathrm{s}}}{P_{\mathrm{s}}/10^{(\mathrm{SNR}/10)} + \mathrm{THD} \cdot P_{\mathrm{s}}}\right). \tag{3.17}$$

3.1.2 Verification: FDC vs. TDC Architecture

The FDC and TDC VCO-based quantizer are both analyzed in the presence of VCO tuning curve nonlinearity. From (3.17), it is shown that the VCO nonlinearity will cause the same distortion for both quantizers. The theoretical SNDR for the FDC VCO-based quantizer is (3.18) and (3.19) for the TDC VCO-based quantizer.

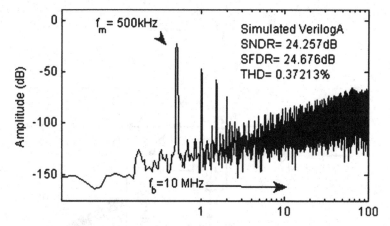

Fig. 3.3 Simulated VerilogA model FDC VCO-based quantizer output including VCO nonlinearity (frequency domain)

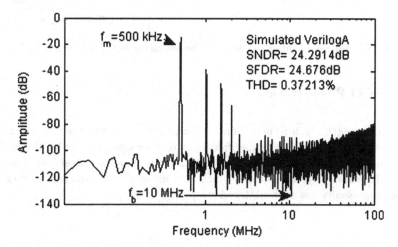

Fig. 3.4 Simulated VerilogA model TDC VCO-based quantizer output including VCO nonlinearity (frequency domain)

$$\text{SNDR}_{\text{FDC}} = 10\log_{10}\left(\frac{1}{1/10^{(60.78/10)} + 0.03721}\right) = 24.29 \text{ dB}, \qquad (3.18)$$

$$\text{SNDR}_{\text{TDC}} = 10\log_{10}\left(\frac{1}{1/10^{(60.68/10)} + 0.03721}\right) = 24.29 \text{ dB}. \qquad (3.19)$$

The VerilogA code for the FDC and TDC VCO-based quantizer is simulated as before in Chap. 2, now with VCO nonlinearity added to the tuning curve. The results are shown in Figs. 3.3 and 3.4 for the FDC and TDC, respectively. The theoretical SNDR closely matches the simulated VerilogA results.

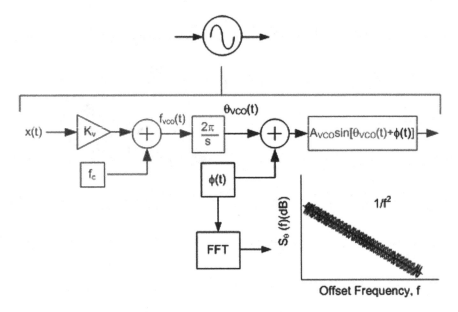

Fig. 3.5 VCO phase noise shown as sideband noise around the VCO carrier frequency

3.2   VCO Phase Noise

Phase noise, $\phi(t)$, creates fluctuations in the phase of the VCO which will cause a skirt around the desired operating frequency (carrier signal) of the VCO, Fig. 3.5. Phase noise is defined in the frequency domain in terms of sideband noise, where the value of phase noise is given at a relative offset frequency from the carrier and usually has a 20 dB/decade slope ($1/f^2$).

3.2.1   Modeling and Analysis

Phase noise in the frequency domain can be represented as jitter in the time domain. Jitter adds to the VCO period which translates to phase noise in the VCO, Fig. 3.6 [1].

To analyze the effects of VCO phase noise, the ring VCO from the previous section is simulated using SpectreRF to provide the phase noise plot in Fig. 3.7. It contains both $1/f^3$ and $1/f^2$ noise.

To analyze the effect of phase noise on the FDC and TDC VCO-based quantizer, phase noise must be added to the ideal VCO equations. Phase noise from the frequency domain must first be converted to time domain jitter, which will add to the instantaneous VCO period. Preexisting models in VerilogA model the $1/f^2$ noise as shown below [1].

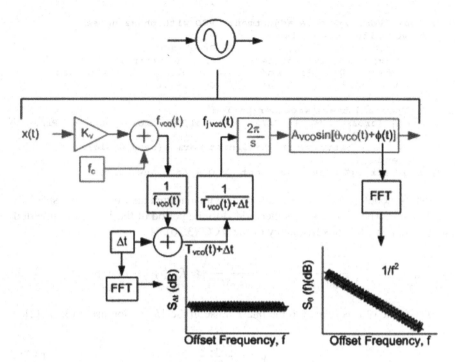

Fig. 3.6 VCO phase noise modeled as time domain jitter

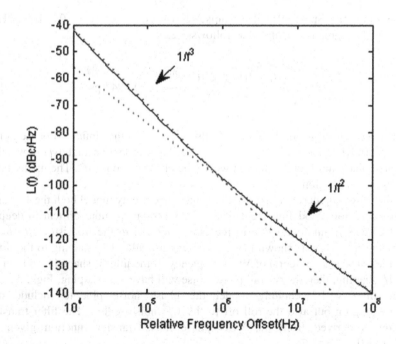

Fig. 3.7 Simulated SpectreRF VCO phase noise containing both third-order and second-order phase noise

```
// Begin VerilogA code adjustment: VCO with phase noise
   // VCO with phase noise
        fvco=Kv*V(in)+fc;       // VCO Frequency
        fjvco=1/(1/fvco+jitter);      // Add jitter
        phase=2.0*`PI*idtmod(fjvco,0,1,-0.5);      // VCO Phase
        vcout=Avco*sin(phase);   // VCO output

   // Update jitter twice per period
        @ (cross(phase + `PI/2, +1, ttol) or cross(phase - `PI/2,
   +1, ttol)) begin
             jitter=sqrt(2)*($rdist_normal(seed,0,sig));
        end
// End VerilogA code adjustment: VCO with phase noise
```

In this model, jitter is considered white, with a normal distribution with standard deviation σ. This jitter value is then added to the period of the VCO, and inverted to get the instantaneous frequency of the VCO (3.20).

$$f_{\text{j-vco}}(t) = \frac{1}{1/f_{\text{vco}}(t) + \Delta t} = \frac{f_{\text{vco}}(t)}{1 + f_{\text{vco}}(t)\Delta t} \approx f_{\text{vco}}(t)[1 - f_{\text{vco}}(t)\Delta t]. \qquad (3.20)$$

The value of σ is related to the phase noise in the $1/f^2$ region using (3.21) [2].

$$\sigma^2 = \frac{f^2 L(f)}{f_c^3}. \qquad (3.21)$$

The simulated SpectreRF phase noise shows $L(f) = -140$ at $f = 100$ MHz. Using (3.21), sigma is calculated as follows(3.22):

$$\sigma = \sqrt{\frac{(100 \text{ MHz})^2 \times 10^{-140/10}}{(500 \text{ MHz})^3}} = 0.8944 \text{ ps.} \qquad (3.22)$$

The comparison of the simulated VerilogA model to the simulated phase noise from SpectreRF is shown in Fig. 3.8. The VerilogA model accurately follows the simulated phase noise in the portion where the phase noise is $1/f^2$. The model falls short in the $1/f^3$ region.

In the following derivations, work is done to accurately model both the $1/f^2$ and $1/f^3$ noise region. Modeling the $1/f^3$ region is becoming more critical in deeply scaled CMOS technologies due to the rapid increase in the transistor $1/f$ noise. Following Fig. 3.6, it is shown that the VCO phase adds $1/f^2$ shaping to the jitter since the phase is the integral of VCO frequency. If the jitter is shaped by a filter to have $1/f$ shaping, then the overall phase noise will have $1/f^3$ shaping, Fig. 3.9.

The filter will accurately model the phase noise plot to include not only $1/f^3$ region but also the roll off to the $1/f^2$ noise region. The filter transfer function is derived from the VCO phase noise transfer function given in Fig. 3.10 [3].

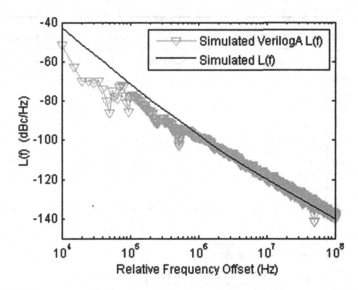

Fig. 3.8 VCO phase noise: VerilogA model vs. simulated SpectreRF

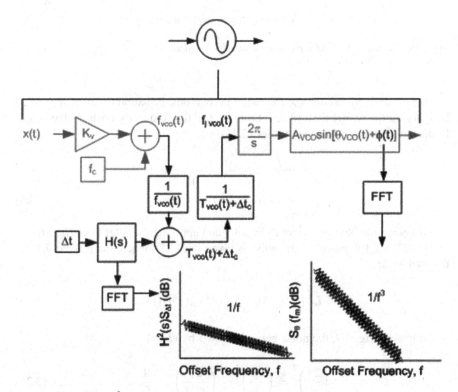

Fig. 3.9 Modeling $1/f^3$ region of phase noise using a filter

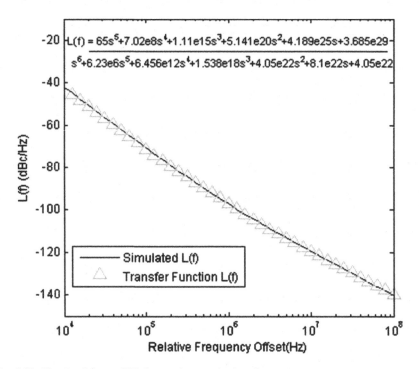

Fig. 3.10 Simulated SpectreRF phase noise transfer function

$H(f)$ is derived through the following derivations. Equation (3.20) is simplified by ignoring the $K_v x(t)$ term (3.23). From (3.23), the frequency contains the center frequency $f_c(t)$ and error due to jitter (3.24).

$$f_{j\_vco}(t) \approx f_c(t)[1 - f_c(t)\Delta t],\tag{3.23}$$

$$f_j(t) \approx f_c^2(t)\Delta t.\tag{3.24}$$

Following the block diagram in Fig. 3.9 and applying the jitter filter $H(f)$ to the jitter in (3.24), the power of the noise is given in (3.25), where the filtered jitter is denoted as Δt_c.

$$L\left\{\left(f_c^2(t)\Delta t_c\right)^2\right\} = f_c^4 S_{\Delta t}(f)H(f).\tag{3.25}$$

Following Fig. 3.9, the noise gets multiplied by (3.26).

$$\left(\frac{2\pi}{s}\right)^2 = \left(\frac{2\pi}{j\omega}\right)^2 = \left(\frac{2\pi}{j2\pi f}\right)^2 = \frac{1}{f^2}.\tag{3.26}$$

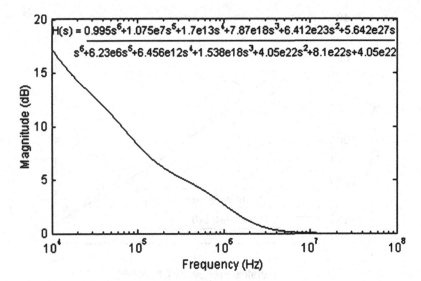

$$H(s) = \frac{0.995s^6 + 1.075e7s^5 + 1.7e13s^4 + 7.87e18s^3 + 6.412e23s^2 + 5.642e27s}{s^6 + 6.23e6s^5 + 6.456e12s^4 + 1.538e18s^3 + 4.05e22s^2 + 8.1e22s + 4.05e22}$$

Fig. 3.11 Jitter filter $H(s)$ PSD and transfer function

Using (3.25) and (3.26), the VCO phase noise, including the jitter filter, is given in (3.27). Note: the same derivation can be followed to relate Δt to σ (3.28).

$$L(f) = f_c^4 S_{\Delta t}(f) H(f) \frac{1}{f^2}, \tag{3.27}$$

$$S_{\Delta t}(f) = \frac{\sigma^2}{f_c}. \tag{3.28}$$

The filter $H(f)$ is derived using (3.27) and (3.28) and Fig. 3.10, ensuring the gain of $H(f)$ is 1 where the value of jitter had previously been derived (3.22). The jitter filter is shown in Fig. 3.11.

This filter is applied to the white noise jitter model and shown below.

```
// Begin VerilogA code adjustment: VCO phase noise filter
  fjvco=1/(1/fvco+jitterc);// Add shaped jitter

  // Shape jitter to get jitterc
  jitterc=laplace_nd(jitter,{0,5.642e27, 6.412e23,7.87e18,
  1.699e13, 1.075e7, 0.995},
  {4.05e22,8.1e22,4.05e22,1.538e18,6.456e12,6.23e6,1});
  // End VerilogA code adjustment: VCO phase noise filter
```

Using the new model, the VCO is simulated using the same value for sigma. The modeled VCO phase noise is plotted against the simulated SpectreRF VCO phase noise and closely matches, Fig. 3.12.

To determine the phase noise effects on the FDC and TDC VCO-based quantizer, (3.23) will be inserted into the previously derived output equations. The FDC VCO-based quantizer output with phase noise can be rewritten as (3.29).

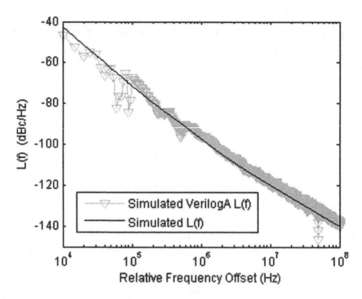

Fig. 3.12 VCO phase noise: VerilogA model (including filter) vs. simulated SpectreRF

$$\mathrm{FDC_{out}}(n) = 2NT_s\left[f_c - f_c^2\Delta t_c + K_v x(n)\right] + \frac{\phi_{\mathrm{vco}}(n-1) - \phi_{\mathrm{vco}}(n)}{\pi}. \qquad (3.29)$$

The power of the phase noise at the FDC output is the integrated noise from the observation time to the system bandwidth (3.30).

$$P_{\mathrm{PN\_FDC}} = \left[2NT_s f_c^2\right]^2 S_{\Delta t}(f) \int_{f_{\mathrm{obs}}}^{f_b} H(f)\mathrm{d}f. \qquad (3.30)$$

The TDC VCO-based quantizer output with phase noise can be rewritten as (3.31), ignoring the phase noise in the quantization error.

$$\mathrm{TDC_{out}}(n) \approx \frac{2\tau_{\mathrm{buff}}\left[f_c - f_c^2\Delta t_c + K_v x(n)\right]}{N} - \frac{4\sqrt{N}\tau_{\mathrm{buff}} f_c^2 q(n)}{N^2}. \qquad (3.31)$$

The power of the phase noise at the TDC output is given as follows (3.32):

$$P_{\mathrm{PN\_TDC}} = \left[\frac{2\tau_{\mathrm{buff}s} f_c^2}{N}\right]^2 S_{\Delta t}(f) \int_{f_{\mathrm{obs}}}^{f_b} H(f)^2\mathrm{d}f. \qquad (3.32)$$

With the bandwidth equal to 10 MHz and the observation time for the simulator equal to 10 kHz, (3.33) is derived using (3.22) and (3.28) and MATLABs "quad" function to assist in solving the integral of $H(f)$.

$$S_{\Delta t}(f) \int_{f_{\mathrm{obs}}}^{f_b} H(f)\mathrm{d}f = \frac{(0.8994\ \mathrm{ps})^2}{500\mathrm{e}6} 1.8876\mathrm{e}7 = 3.02\mathrm{e}^{-26}. \qquad (3.33)$$

3.2.2 Verification: FDC vs. TDC Architecture

The SNR equation is now written as the power of the signal divided by the power of the noise plus the power of the phase noise (3.34).

$$\text{SNR} = 10\log_{10}\left(\frac{P_\text{s}}{P_\text{N} + P_\text{PN}}\right). \tag{3.34}$$

The SNR for the FDC VCO-based quantizer with phase noise is given in (3.35) and its simplified form is given in (3.36).

$$\text{SNR}_{\text{FDC}} = 10\log_{10}\left(\frac{\left(2NT_\text{s}K_\text{v}A_\text{m}/\sqrt{2}\right)^2}{8(\pi^2 f_\text{b}^3/36 f_\text{s}^3) + \left[2NT_\text{s}f_\text{c}^2\right]^2 S_{\Delta t}(f)\int_{f_\text{obs}}^{f_\text{b}} H(f)df}\right). \tag{3.35}$$

$$\text{SNR}_{\text{FDC}} = 10\log_{10}\left(\frac{9 f_\text{s}(NK_\text{v}A_\text{m})^2}{\pi^2 f_\text{b}^3 + 18 N^2 f_\text{s} f_\text{c}^4 S_{\Delta t}(f)\int_{f_\text{obs}}^{f_\text{b}} H(f)df}\right) = 59.42 \text{ dB}. \tag{3.36}$$

The SNR for the TDC VCO-based quantizer with phase noise is given in (3.37) and its simplified form is given in (3.38).

$$\text{SNR}_{\text{TDC}} = 10\log_{10}\left(\frac{\left(2\tau_{\text{buff}}K_\text{v}A_\text{m}/N\sqrt{2}\right)^2}{(32 f_\text{b}\tau_{\text{buff}}^4 f_\text{c}^4)/(N^4 12 f_\text{s}) + ((2\tau_{\text{buff}}f_\text{c}^2)/N)^2 S_{\Delta t}(f)\int_{f_\text{obs}}^{f_\text{b}} H(f)^2 df}\right), \tag{3.37}$$

$$\text{SNR}_{\text{TDC}} = 10\log_{10}\left(\frac{3 f_\text{s}(NK_\text{v}A_\text{m})^2}{4 f_\text{b}\tau_{\text{buff}}^2 f_\text{c}^4 + 6 N^2 f_\text{s} f_\text{c}^4 S_{\Delta t}(f)\int_{f_\text{obs}}^{f_\text{b}} H(f)^2 df}\right)$$

$$= 60.33 \text{ dB}. \tag{3.38}$$

The FDC and TDC are simulated as before in Chap. 2, only this time with phase noise. The PSD of the FDC and TDC VCO-based quantizer output is shown in Figs. 3.13 and 3.14 for the FDC and TDC, respectively. The simulated VerilogA model SNR closely matches the theoretical value.

Both the FDC and TDC SNR degrade by about the same factor. The power of the phase noise in both cases gets multiplied by the number of stages, sampling frequency, and center frequency.

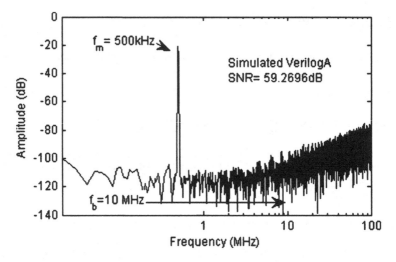

Fig. 3.13 Simulated VerilogA model FDC VCO-based quantizer output including VCO phase noise (frequency domain)

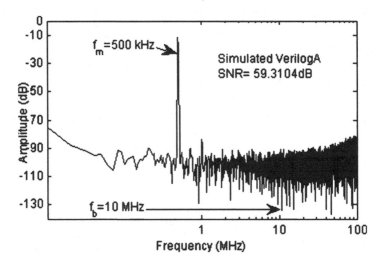

Fig. 3.14 Simulated VerilogA model TDC VCO-based quantizer output including VCO phase noise (frequency domain)

3.3 Sampling Clock Jitter

Jitter in the sampling clock leads to sampling uncertainty. In a typical ADC, a jittery sampling clock will cause a skirt around the desired output signal, Fig. 3.15. This is also the case with the VCO-based quantizer. The FDC will have further error caused by integration error [4].

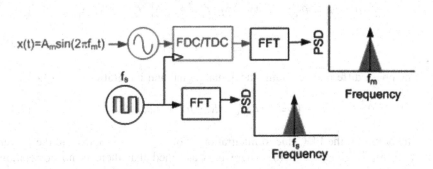

Fig. 3.15 Block diagram: sampling clock jitter

3.3.1 Modeling and Analysis

To analyze the effects of sampling clock jitter, jitter will be modeled using the VerilogA code from the previous section, assuming only second-order noise. The sampling clock frequency is 2 GHz, a reasonable period jitter for this clock is $\sigma = 2$ ps rms [5]. As stated before, sampling the input signal with a jittery clock will cause a skirt around the signal at the output spectrum. This is known as sampling error. Error due to sampling was not observed during simulation and is ignored in this case. However, the FDC will have additional error due to integration which is quite substantial and must be modeled. A similar derivation is followed in ref. [4]. Consider the FDC output equation in (3.39)

$$\text{FDC}_{\text{out}}(n) = \frac{N}{\pi} [\theta_{\text{vco}}(n) - \theta_{\text{vco}}(n-1) + \phi_{\text{vco}}(n-1) - \phi_{\text{vco}}(n)]. \quad (3.39)$$

The sampled VCO phase is given in (3.40).

$$\theta_{\text{vco}}(n) = 2\pi \int_{(n-1)T_s}^{nT_s} f_c + K_v x(\tau) d\tau. \quad (3.40)$$

Absolute jitter, $\tau_{\text{aj}}(n)$, will cause the sampled phase to become (3.41), where the time reference for integration has been shifted.

$$\theta_{\text{vco}}(n) = 2\pi \int_{T_s(n-1)+\tau_{\text{aj}}(n-1)}^{T_s(n)+\tau_{\text{aj}}(n)} f_c + K_v x(\tau) d\tau$$

$$= 2\pi \sum_{i=1}^{n} [T_s + \tau_{\text{aj}}(i)] [f_c + K_v x(i)]. \quad (3.41)$$

The phase difference with clock jitter is shown in (3.42) where absolute jitter is replaced by period jitter, τ_{pj}, due to the relationship in (3.43)

$$\theta_{\text{vco}}(n) - \theta_{\text{vco}}(n-1) = 2\pi \left[T_s + \tau_{\text{pj}}(n) \right] \left[f_c + K_v x(n) \right], \tag{3.42}$$

$$\tau_{\text{aj}}(n) - \tau_{\text{aj}}(n-1) = \tau_{\text{pj}}(n). \tag{3.43}$$

The phase difference contains the actual signal and integration error (3.44).

$$\theta_{\text{vco}}(n) - \theta_{\text{vco}}(n-1) = 2\pi T_s [f_c + K_v x(n)] + 2\pi \tau_{\text{pj}} [f_c + K_v x(n)]. \tag{3.44}$$

The output of the FDC due to integration error is given in (3.45) and the power of the error is given in (3.46) where it is assumed that there is no correlation between the two error components [4]:

$$\text{FDC}_{\text{out\_ie}}(n) = 2\tau_{\text{pj}} [f_c + K_v x(n)], \tag{3.45}$$

$$P_{\text{IE}} = 4 S \tau_{\text{pj}} \left[f_c^2 + \left(\frac{K_v A_m}{\sqrt{2}} \right)^2 \right]. \tag{3.46}$$

The power of jitter is given in (3.47) [4].

$$S \tau_{\text{pj}} = \sigma^2 = (2 \text{ ps})^2. \tag{3.47}$$

3.3.2 Verification: FDC vs. TDC Architecture

The SNR equation is now written as the power of the signal divided by the power of the noise plus the power of the integration error (3.48).

$$\text{SNR} = 10\log_{10} \left(\frac{P_s}{P_N + P_{\text{IE}}} \right). \tag{3.48}$$

The SNR for the FDC VCO-based quantizer with integration error is given in (3.49) and its simplified form is given in (3.50).

$$\text{SNR}_{\text{FDC}} = 10\log_{10} \left(\frac{\left((2NT_s K_v A_m) / \sqrt{2} \right)^2}{8 (\pi^2 f_b^3 / 36 f_s^3) + 4\tau_{\text{pj}}^2 \left[f_c^2 + (K_v A_m / \sqrt{2})^2 \right]} \right), \tag{3.49}$$

$$\text{SNR}_{\text{FDC}} = 10\log_{10} \left(\frac{9 f_s (N K_v A_m)^2}{\pi^2 f_b^3 + 18 f_s^3 \tau_{\text{pj}}^2 \left[f_c^2 + (K_v A_m / \sqrt{2})^2 \right]} \right) = 48 \text{ dB}. \tag{3.50}$$

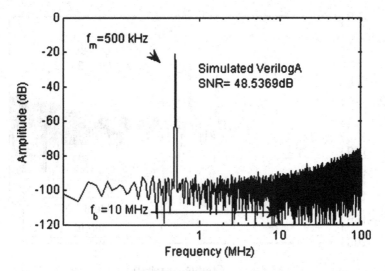

Fig. 3.16 Simulated VerilogA model FDC VCO-based quantizer output including clock jitter (frequency domain)

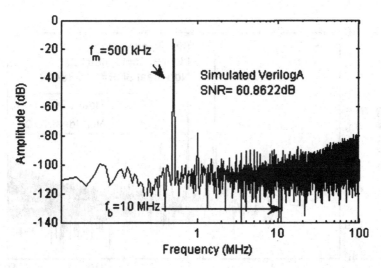

Fig. 3.17 Simulated VerilogA model TDC VCO-based quantizer output including clock jitter (frequency domain)

The FDC and TDC are simulated as before in Chap. 2, only this time with clock jitter. The PSD of the FDC and TDC VCO-based quantizer is shown in Figs. 3.16 and 3.17 for the FDC and TDC, respectively. The simulated VerilogA model SNR closely matches the theoretical value.

The FDC SNR degrades significantly due to sampling clock jitter when compared with the TDC. This is due to the fact that the sampling clock jitter also causes integration error of the VCO phase which destroys the first-order noise shaping of the FDC.

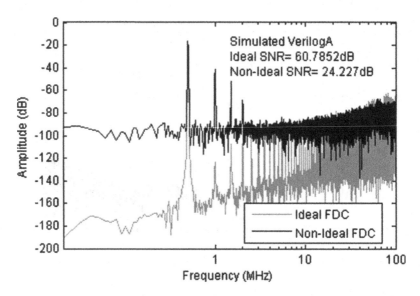

Fig. 3.18 PSD FDC VCO-based quantizer ideal vs. nonideal output

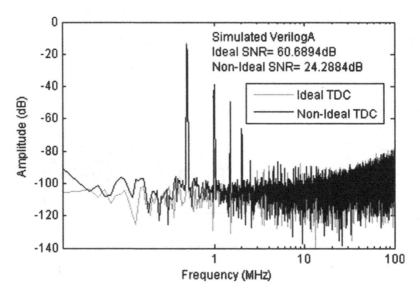

Fig. 3.19 PSD TDC VCO-based quantizer ideal vs. nonideal output

3.4 Summary

Nonidealities of the VCO-based quantizer were analyzed to compare the performance of the FDC vs. the TDC approach. Both quantizers degrade similarly due to VCO nonlinearity and phase noise. In the presence of sampling clock jitter, the FDC

Table 3.1 Effects of nonidealities on FDC/TDC VCO-based quantizer SNR

	FDC SNR		TDC SNR	
	Theoretical (dB)	VerilogA (dB)	Theoretical (dB)	VerilogA (dB)
Ideal	60.11	60.78	61.03	60.68
VCO nonlinearity	24.29	24.25	24.29	24.29
VCO phase noise	58.63	59.27	59.19	59.45
Sampling clock jitter	48	48.53	–	60.86
All nonidealities	24.22	24.21	24.29	24.28

resolution decreased due to destruction of the first-order noise shaping characteristic. The TDC is far less sensitive to sampling clock jitter which had no effect on the resolution. All the nonidealities are modeled and the results are shown in Figs. 3.18 and 3.19 for the FDC and TDC, respectively.

As shown, the first-order noise shaping of the FDC is destroyed. The nonlinearity of the VCO tuning curve is the bottleneck to achieving high SNR and dominates over the other nonidealities in both quantizers.

The SNR for the FDC and TDC VCO-based quantizer is summarized in Table 3.1. The theoretical and modeled SNR with nonidealities is shown. The theoretical values closely match the modeled values and have been confirmed with multiple simulations.

References

1. Takahashi M, Ogawa K, Kundert KS (1999) VCO jitter simulation and its comparison with measurement. In: Proceedings of the ASP-DAC 1999 Asia and South Pacific Design Automation Conference, 1999, Wanchai, 18–21 Jan 1999, pp 85–88
2. Poore R (2001) Phase noise and jitter. In: Agilent EEsof EDA. Overview of phase noise and jitter. Agilent Technologies, Inc.
3. Henrik F, Martin (2004) JITTER MANAGING 26/5-04. Electronic Devices, Department of Electrical Engineering, Linköping University, Sweden
4. Kim J et al (2010) Analysis and design of voltage-controlled oscillator based analog-to-digital converter. IEEE Trans Circ Syst I: Reg Pap 57(1):18–30
5. Farjad-Rad R et al (2002) A 0.2-2 GHz 12 mW multiplying DLL for low-jitter clock synthesis in highly-integrated data communication chips. In: 2002 IEEE International Solid-State Circuits Conference (ISSCC), Digest of Technical Papers, San Francisco, CA, 3–7 Feb 2002, vol 1, pp 76–77

Chapter 4
Further Analyses: FDC vs. TDC

The following sections provide some final thoughts and analysis of the FDC and TDC VCO-based quantizer, such as the circuit requirements for each quantizer and performance of the quantizers, given a two-tone test.

4.1 FDC/TDC Circuit Requirements

The circuit requirements for each quantizer is quite different, depending on the application one quantizer might be advantageous over the other. One requirement for both architectures is the VCO operating frequency must be much greater than the input signal bandwidth, to ensure proper transfer of data.

The circuit requirements for the FDC are as follows: the sampling frequency must be 2× the maximum VCO frequency, according to the Nyquist Criteria. Since the maximum VCO frequency will be much greater than the signal frequency, this system will always be oversampled. The quantizer resolution increases with: VCO tuning range, multiphase outputs, and sampling frequency to system bandwidth. It is desirable to have a multiphase VCO with large tuning range to increase the resolution. The multiphase VCO is implemented with a ring VCO, increasing the number of stages or phase taps leads to a lower operation frequency or increase in current consumption (4.1). Requirements of a large tuning range may also lead to higher VCO phase noise and tuning nonlinearity [1].

$$f_{osc} \propto \frac{I_{dd}}{2N}.$$

(4.1)

There is also a requirement to have a zero dc offset at the FDC output [2]. This means that there are an equal number of 1's and 0's when the VCO frequency is operating at its free running center frequency. For this to occur, the center frequency must be 4× smaller than the sampling frequency. This becomes important when the VCO-ADC is used in a $\Delta\Sigma$ modulator.

S. Yoder et al., *VCO-Based Quantizers Using Frequency-to-Digital and Time-to-Digital Converters*, SpringerBriefs in Electrical and Computer Engineering, DOI 10.1007/978-1-4419-9722-7_4, © Springer Science+Business Media, LLC 2011

Table 4.1 FDC/TDC circuit requirements

Restrictions	FDC	TDC
f_s, sampling frequency	$f_s \geq 2 \times \text{VCO}_{f_{max}}$	$f_s \geq 2 \times f_b$
VCO frequency	$f_c = f_s/4$	$\text{VCO}_{T_{max}} \leq 1.5 \times \text{Total delay}$
		$\text{VCO}_{T_{min}} \geq t_{buff}$
VCO	Ring VCO for MB operation	Any
SNR tuning knobs	$\uparrow(f_s, N, K_v)$	$\uparrow(f_s, N, K_v)$
		$\downarrow(t_{buff}, f_c)$

The circuit requirements for the TDC are as follows: the sampling frequency must be $2\times$ the input frequency to the VCO. The delay chain must capture an entire VCO pulse. This requires the maximum VCO period to be 1.5 times smaller than the total delay of the chain. Also the minimum VCO period should not be less than a buffer delay. Since the TDC relies on a single output of the VCO, it may be implemented with a RVCO or LC tank.

The FDC/TDC circuit requirements are summarized in Table 4.1. Also the SNR tuning knobs for each architecture are outlined; the TDC has two additional design criteria to enhance the SNR.

4.2   Two-Tone Test

It is also important to determine how the ideal quantizers react to a two-tone test. The output of the TDC period is inverted to give frequency; nonlinear relationships exist and might cause some distortion. The two-tone test for both quantizers is tested using two sinusoids each with the same amplitude. The two frequencies of this test will be $f_m = 500$ and 125 kHz. The following first-order and second-order distortion products might exist (4.2)–(4.4).

$$500 \text{ kHz} \pm 125 \text{ kHz} = 375 \text{ kHz}, \ 625 \text{ kHz}, \tag{4.2}$$

$$2(500 \text{ kHz}) \pm 125 \text{ kHz} = 875 \text{ kHz}, \ 1.125 \text{ MHz}, \tag{4.3}$$

$$2(125 \text{ kHz}) \pm 500 \text{ kHz} = 25 \text{ kHz}, \ 75 \text{ kHz}. \tag{4.4}$$

The two-tone test results are shown in Figs. 4.1 and 4.2 for the FDC and TDC, respectively. The FDC does not experience intermodulation distortion, while the TDC does. The TDC output spectrum contains many spurs which degrade the SNR. This can be attributed to the fact that the TDC measures the period of the signal and inverts to get frequency which is a nonlinear relationship. The TDC also has a nonlinear relationship of its quantization noise with the input signal which has been ignored in previous analysis.

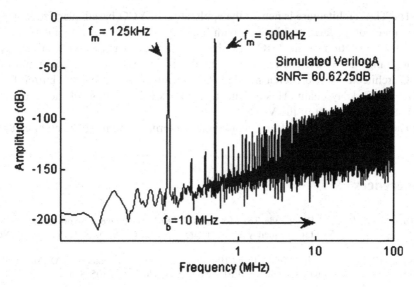

Fig. 4.1 Two-tone test of a FDC VCO-based quantizer (frequency domain)

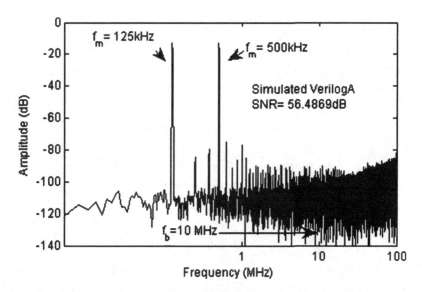

Fig. 4.2 Two-tone test of a TDC VCO-based quantizer (frequency domain)

4.3 Final Thoughts

The FDC architecture offers simplicity and inherent first-order noise shaping. The resolution of the FDC VCO-based quantizer increases with larger VCO tuning range, phase outputs of the VCO, and OSR. The resolution is highly dependent on the VCO. In Chap. 3, it is shown that the FDC architecture is sensitive to clock jitter.

The TDC architecture is new to the application of VCO-based quantizers. It can be a quite complicated architecture with linearity issues as shown by the two-tone test. The resolution of the TDC VCO-based quantizer increases with larger VCO tuning range and OSR. The resolution can also be increased by cascading multiple TDC architectures, and decreasing the VCO operating frequency, and buffer delay of the TDC delay chain. This architecture allows enhancement of the resolution which is independent of the VCO.

The TDC architecture offers a comparable alternative to the FDC architecture.

References

1. Tuan-Vu C, Wisland DT, Lande TS, Moradi F (2008) Low-voltage, low-power, and wide-tuning-range ring-VCO for frequency $\Delta\Sigma$ modulator. In: NORCHIP, 2008, Tallinn, 16–17 Nov 2008, pp 79–84
2. Straayer MZ, Perrott MH (2008) A 12-bit 10-MHz bandwidth, continuous-time $\Delta\Sigma$ ADC with a 5-bit 950-MSs VCO-based quantizer. IEEE J Solid-State Circ 43(4):805–814

Chapter 5
Conclusions

In summary, this document presented an analytical and modeling approach to understanding the VCO-based quantizer. The VCO-based quantizer architecture consists of a VCO and digital time quantizer to quantize the VCO output signal. Two digital time quantizers were modeled and analyzed; one consists of an FDC and the other used a TDC. The FDC architecture has been widely used due to its inherent first-order noise shaping characteristic. The TDC architecture is applied to the VCO-based quantizer here for the first time. Theoretical modeling and analysis of the two quantizers was performed to determine the resolution, and verified through a VerilogA model.

Comparison of these quantizers in the presence of circuit nonidealities such as VCO nonlinearity, phase noise, and sampling clock jitter was performed. Theoretical modeling and analysis of nonidealities were consistent with the VerilogA model and showed several interesting results. VCO nonlinearity remains the bottleneck of these types of quantizers and has been remedied by several methods such as use of a $\Delta\Sigma$ modulator or linearization of the VCO tuning curve. While both quantizers responded similarly to VCO nonlinearity and phase noise, the TDC was less sensitive to sampling clock jitter. This is due to the fact that the FDC quantizes the VCO phase which is the integral of VCO frequency. Sampling clock jitter introduces integration error which corrupts the first-order noise shaping of the FDC.

The two architectures have slightly different circuit requirements and one architecture might be advantageous over the other depending on its application. The FDC architecture is very simple, while the TDC architecture is quite complex and not well understood since its application to the VCO-based quantizer is new. The TDC showed comparable performance to the FDC architecture but suffers from linearity issues.

S. Yoder et al., *VCO-Based Quantizers Using Frequency-to-Digital and Time-to-Digital*   55
Converters, SpringerBriefs in Electrical and Computer Engineering,
DOI 10.1007/978-1-4419-9722-7_5, © Springer Science+Business Media, LLC 2011